International Environmental Labelling

VOL.2 OF 11

For All People who wish to take care of Climate Change

Energy & Electrical Industries: (Renewable Energy, Biofuels, Solar Heating & Cooling, Hydroelectric Power, Solar Power, Wind Power, Energy Conservation, Geothermal and Nuclear Power)

Jahangir Asadi
Vancouver, BC CANADA

Suggest an ecolabel

If you think that we missed a label and/or you are an ecolabelling body, please consider to submit for the next editions of our 11 Volumes International Eco-labelling Book series. Please send your details, and we'll review your suggestions. Our goal is to be as comprehensive as possible, so thank you for your help!
info@TopTenAward.Net

Copyright © 2022 by Top Ten Award International Network.

All rights reserved. No part of this publication may be reproduced, distributed or transmitted in any form or by any means, including photocopying, recording, or other electronic or mechanical methods, without the prior written permission of the publisher, except in the case of brief quotations embodied in critical reviews and certain other noncommercial uses permitted by copyright law. For permission requests, write to the publisher, addressed "Attention: Permissions Coordinator," at the address below.

Published by: Top Ten Award International Network
Vancouver, BC **CANADA**
Email: Info@TopTenAward.net
www.TopTenAward.net

Ordering Information:
Quantity sales. Special discounts are available on quantity purchases by universities, schools, corporations, associations, and others. For details, contact the "Sales Department" at the above mentioned email address.

International Environmental Labelling Vol.2/J.Asadi—2nd ed.
ISBN 978-1-7773356-4-9

Contents

About TTAIN 10
Introduction 13
General principles of environmental labelling 20
Types of environmental labelling 24
Types I environmental labelling 28
Types II environmental labelling 42
Types III environmental labelling 48
What is Renewable Energy? 52
The future of Renewable Energy 63
TTAIN environmental pioneers 68
International Organizations 72
Bibliography 75
Search by logos 82
A sample of Biofuel 100
Environmental friendly photos 102

I dedicate this book to my dear kids: Tara, Tarannom & Taha

Acknowledgements:

I wish to thank my committee members, who were more than generous with their expertise and precious time. I would like to acknowledge and thank the Top Ten Award International Network for allowing me to conduct my research and providing any assistance requested.

It should be noted that all the required permissions for using the logos and trade marks has been obtained to be published in this volume.

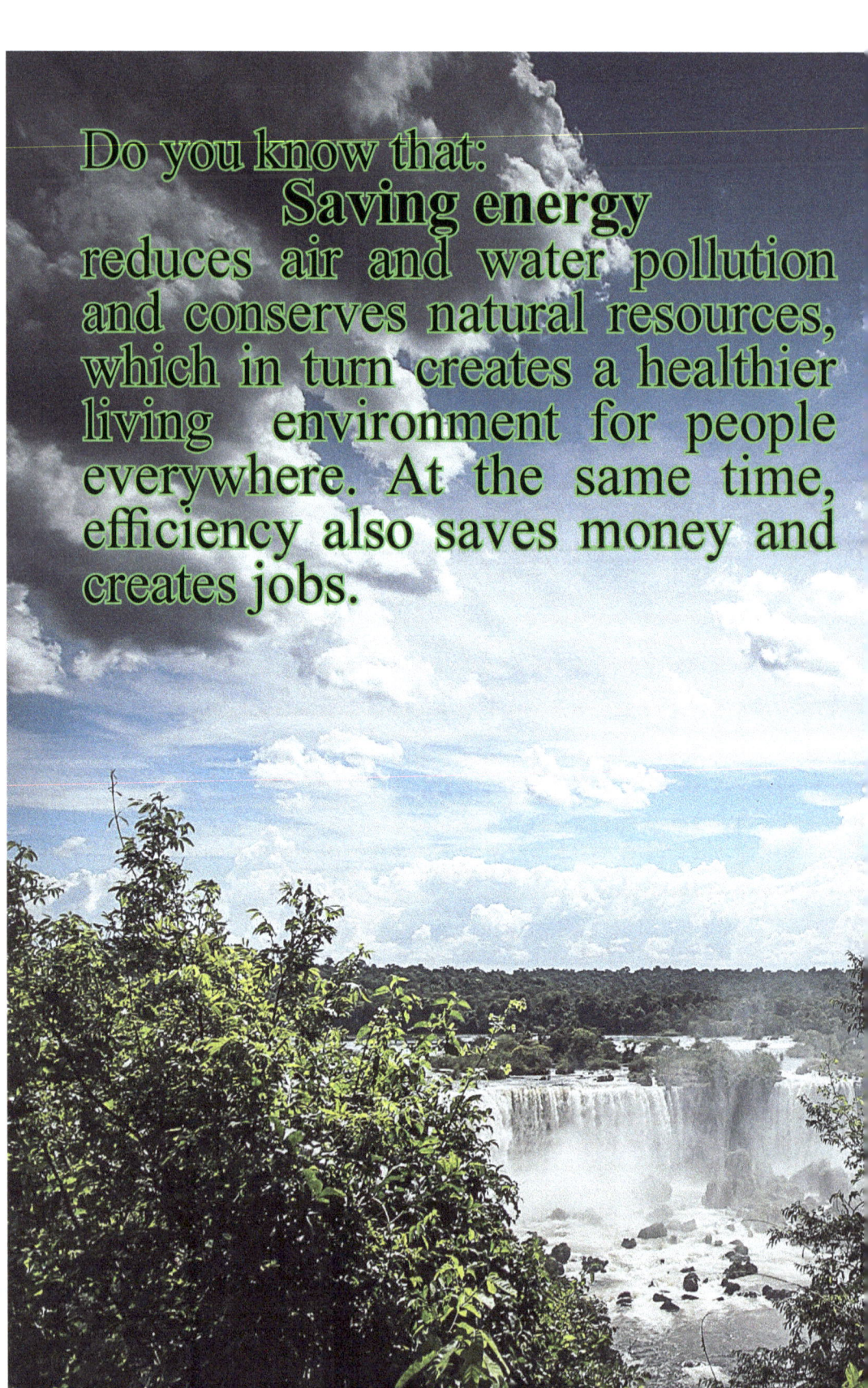

Do you know that:
Saving energy reduces air and water pollution and conserves natural resources, which in turn creates a healthier living environment for people everywhere. At the same time, efficiency **also saves money and creates jobs.**

About TTAIN

Top Ten Award International Network

Top Ten Award international Network (TTAIN) was established in 2012 to recognize outstanding individuals, groups, companies, organizations representing the best in the public works profession.

TTAIN publishing books related to international Eco-labeling plans to increase public knowledge in purchasing based on the environmental impacts of products.

Top Ten Award International Network provides A to Z book publishing services and distribution to over 39,000 booksellers worldwide, including Apple, Amazon, Barnes & Noble, Indigo, Google Play Books, and many more.

Our services including: editing, design, distribution, marketing
TTAIN Book publishing are in the following categories:

Student
Standard
Business
Professional
Honorary

We focus on quality, environmental & food safety management systems , as well as environmnetal sustain for future kids. TTAIN also provide complete consulting services for QMS, EMS, FSMS, HACCP and Ecolabeling based on international standards.

ISO 14024 establishes the principles and procedures for developing Type I environmental labelling programmes, including the selection of product categories, product environmental criteria and product function characteristics, and for assessing and demonstrating compliance. ISO 14024 also establishes the certification procedures for awarding the label.

TTAIN has enough experiences to help create new ecolabeling programmes in different countries all over the world.
For more detail visit our website : http://toptenaward.net
and/or send your enquiery to the following email:
info@toptenaward.net

CHAPTER 1

Introduction

This book is dedicated to the subject of environmental labels. The basis for the classification of its parts goes back to the types of environmental labelling according to the classifications provided by the International Organization for Standardization. In each section, while presenting the relevant definitions, I mention the existing international standards and present examples related to each type of labelling. Environmental labelling is an important and significant topic, and its richness is added to every day, which has attracted the attention of many experts and researchers around the world. The idea of compiling this book, came to my mind when I observed that national environmental labelling models have been developed in most countries of the world, but in many other countries, the initial steps have not been taken yet. Therefore, I decided to create the first spark for the development of environmental labelling patterns in other countries by collecting appropriate materials and inserting samples of labelling patterns of different countries of the world. It should be noted that the description of each environmental label in this book does not indicate their approval or denial; they are included only to increase the awareness of all enthusiasts and consumers of the meanings and concepts derived from such labels. We hereby ask all interested parties around the world who wish to start an environmental labelling program in their country to benefit from our intellectual assistance and support in the form of consulting contracts. Increasing human awareness of the urgent need to protect the environment has led to changes in all levels of activities, including the production of marketing products, consumption, use, and sale of goods and services at the national and international levels. Stakeholders involved in environmental protection include consumers, producers, traders, scientific and technological institutes, national authorities, local and international organizations, environmental gatherings, and human society in general. Decisions by consumers and sellers of products are made not only on the basis of key points such as quality, price, and availability of

products but also on the environmental consequences of products, including the consequences that a product can have before, after and during production. The most important environmental consequences include water, soil, and air pollution along with waste generation, especially hazardous waste. Further consequences include noise, odor, dust, vibration, and heat dissipation as well as energy consumption using water, land, fuel, wood, and other natural resources. There are further effects on certain parts of the ecosystem and the environment. In addition, the environmental consequences not only include the natural use of the products but also abnormal and even emergency or accidental uses. The basis of studies and studies in this field is done through product life cycle evaluation, which generally involves the study and evaluation of environmental aspects and consequences of a category (product, service, etc.) because of the preparation of raw materials for production until they are used or discarded. Sometimes the phrase "review from cradle to grave" is used for such an evaluation. In addition to the above, the environmental consequences that may occur at any stage of the product life cycle, including the preliminary stages and its preparation, production, distribution, operation, and sale, should also be considered when evaluating it. This type of evaluation refers to product life cycle analysis from an environmental point of view,"

which is a useful tool for measuring the degree of environmental health of a product, comparing different products, improving product quality, and confirming the environmental health claims of the product. The environmental health analysis tool for products and services facilitates their placement in domestic or foreign markets, considering that the awareness of consumers and retailers about the environmental consequences of the product has increased, as has the accurate and explicit measurement by the people in charge at all levels. Local, national, and international in the field of environmental protection. Products that can claim to be environmentally complete in all stages of their life cycle and meet the mandatory and optional environmental needs are considered successful products. Environmental messages refer to the policies, goals, and skills of product manufacturing companies as part of the environmental management systems in which they are applied, and consumers and retailers are increasingly paying attention to this issue when making purchasing decisions. In addition, companies have been encouraged and even forced to adapt their environmental management systems to agencies and retailers and to local, national, international, and other environmental issues.

The environmental health message of a product can be conveyed to the consumer in various ways, including implicitly or explicitly. For example, the implicit or implicit message conveyed directly by the product to the customer is that the product is suitable for the intended use and purpose, and, without material waste in size, weight, and dimensions, is perfectly proportioned and without additional packaging. Sometimes it is necessary to convey these messages and claims about the correctness of the product quite clearly through magazines or other media as well as through certificates that are accurate, simple, and convincing to the consumer in the form of a label. These messages must be accurate and fact-based; otherwise they will nullify the product and create contradictory effects. Confirmation of these claims by a third-party organization will increase its credibility. It should also be noted that the multiplicity of these messages, depending on the type of products or companies producing them, confuses consumers in the market and also creates artificial boundaries or causes a differentiated distinction against certain products or companies. Various models, principles, and methods have been provided by local, regional, national, and international organizations to demonstrate product life cycle analysis and other guidelines on environmental management systems and their labels. At the national level, significant advances have been made in the design of environmental labels in various countries, including developing countries and the Scandinavian countries. For example, the first project was designated in Germany as a Blue Angel in 1977, later on Canada in 1988, the Scandinavian countries and Japan in 1989, the United States and New Zealand in 1990, India, Austria, and Australia in 1991, And in 1992, Singapore, the Republic of Korea, and the Netherlands developed their national environmental labelling. Environmental labels are an environmental management tool that is the subject of a series of ISO 14000 standards. These environmental labels provide information about a product or commodity in terms of its broad environmental characteristics, whether it is about a specific environmental issue or about other characteristics and topics.Interested and pro-environmental buyers can use this information when choosing products or goods. Product makers with these environmental labels hope to influence people's purchasing decisions. If these environmental labels have this effect, the share of the product in question can increase, and other suppliers may create healthy environmental competition by improving the environmental aspects of their products and commodities. The overall goal of environmental labels is to convey acceptable and accurate information that is in no way misleading regarding the environmental aspects of products and commodities, and they encourage the consumer to buy and produce products that reduce stress on the environment. Environmental labelling must follow the general principles that the International Organization for Standardization has published in a collection entitled the ISO 14020 standard, which refers to these general principles here. It should be noted that other documents and laws in this field are considered if they are in accordance with the principles set out in ISO 14020.

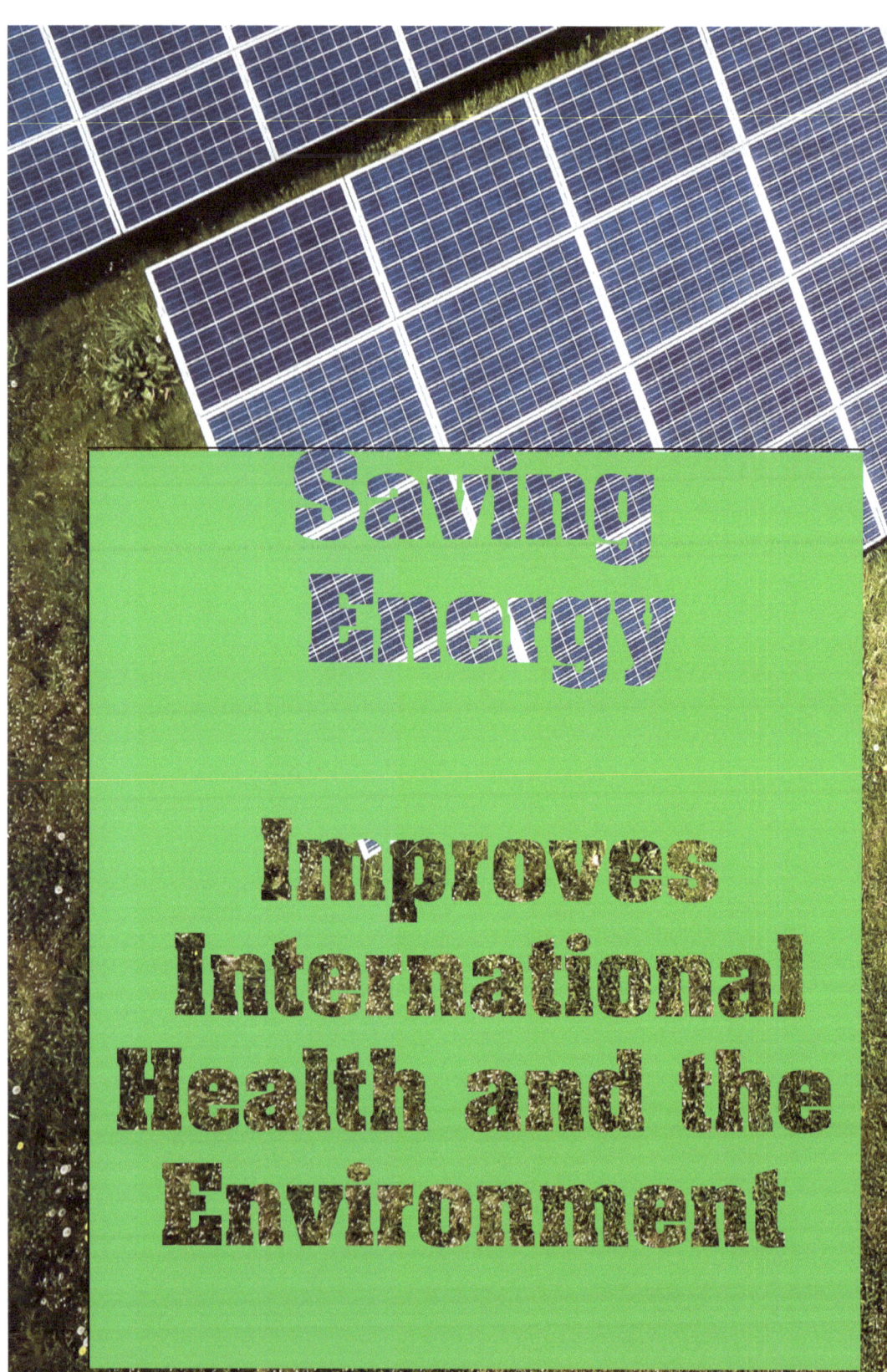

CHAPTER 2

General Principles on Environmental Labelling

1 The First Principle: Evironmental notices and labels must be accurate, verifiable, relevant, and in no way misleading and/or deceptive.

2 The Second Principle: Procedures and requirements for environmental labels will not be ready for selection unless they are implemented by affecting or eliminating unnecessary barriers to international trade.

3 The Third Principle: Environmental notices and labels will be based on scientific analysis that is sufficiently broad and comprehensive, and to support this claim, the product must be reliable and reproducible.

4 The Fourth Principle: The process, methodology, and any criteria required to support the announcements on environmental labels will be available upon request all interested groups.

5 The Fifth Principle: Development and improvement of environmental notices and labels should be considered in all aspects related to the service life of the product.

6 The Sixth Principle: Announcements on environmental labels will not prevent initiative and innovation but will be important in maintaining environmental implementation.

7 The Seventh Principle: Any enforcement request or information requirement related to environmental notices and labels should be limited to the necessary information to establish compliance with an acceptable standard and based on the notification standards and environmental labels.

8 The Eighth Principle: The process of improving the announcement and environmental labels should be done by an open solution with interested groups. Reasonable impressions must be made to reach a consensus through this process.

9 The Ninth Principle: Information on the environmental aspects of the product and goods related to an advertisement and environmental label will be prepared for buyers and interested buyers from a group consisting of an advertisement and an environmental label.

CHAPTER 3

Types of Environmental Labelling

At present, according to the classification provided by the International Organization for Standardization, there are three types of environmental labelling patterns:

1. Type I labelling: This labelling is known as eco-labelling, and because it is difficult to translate this word into many languages, it presents another reason to adhere to a numerical classification system. In the content of Type I labelling, a set of social commitments that creates criteria according to the scientific principles on the basis of which a product is environmentally preferable is discussed. Consumers are then instructed in assessing environmental claims and must decide which packaging is more important.

2. Type II labelling: refers to the claims made on product labels in connection with business centers. This includes familiar claims such as recyclable, ozone-friendly, 60% phosphate-free, and the like. This type of labelling can be in the form of a mark or sentence on the product packaging. Some of them are valid environmental claims—and some can be completely misleading. Usually, all countries have laws against deceptive advertisements, so why has the International Organization for Standardization discussed this issue? The answer is that it is not clear whether the environmental claims have a technical basis or whether the ad is meaningless.

3 Type III labelling: is a distinct form of third-party environmental labelling pattern designed to avoid the difficulties that can result from type-one labelling. Technical committee for Environment of International organization for Standardization has undertaken a new project to standardize guidelines and Type III labelling methods. One of the main objections raised by industries to Type I labelling is the basis for its management.

Problems with the current energy system:

- Energy and climate change
- Climate change is already happening
- Energy access and energy poverty
- Global inequalities in energy use
- Political insecurity, corruption and conflict
- Energy waste

CHAPTER 4

Type I Environmental Labelling

Type I labelling: This labelling is known as eco-labelling, and because it is difficult to translate this word into many languages, it presents another reason to adhere to a numerical classification system. In the content of Type I labelling, a set of social commitments that creates criteria according to the scientific principles on the basis of which a product is environmentally preferable is discussed. Consumers are then instructed in assessing environmental claims and must decide which packaging is more important.

Type I adhesive has the following specifications:
A. Has an optional third-party template.
B. When the product meets a certain standard, the labelling of this product is included.
C. The purpose of this program is to identify and promote products that play a pioneering role in terms of environment, which means its criteria are at a higher level than the average environmental performance.
D. Acceptance/rejection criteria are determined for each group of products and are publicly available.
E. The criteria are adjusted after considering the environmental consequences of the product life cycle.

Examples of Type I Labelling:
In this section, and considering the importance of this type of labelling, I provide a description of some examples of Type I labelling related to some countries along with a list of products on which this mark is placed.

Finland

EKOenergy is a global, nonprofit ecolabel for renewable energy (electricity, gas, heat and cold). In addition to being renewable, the energy sold with the EKOenergy label fulfils sustainability criteria and helps finance projects that combat energy poverty. The financed projects address several SDGs and give the opportunity for consumers to achieve more with their purchase. The EKOenergy label can be combined with all sourcing methods. EKOenergy-labelled energy is currently available in 40+ countries.

Users of EKOenergy-labelled energy can use the logo on products made with EKOenergy. By using our internationally recognised logo, individuals and companies demonstrate their commitment to renewables.

Contact info:
Steven Vanholme,
steven.vanholme@sll.fi

Singapore

Established in 1995, the Singapore Environment Council (SEC) is an independently managed, non-profit and non-governmental organisation (NGO). As Singapore's first United Nations Environment Programme (UNEP)-accredited environmental NGO, we influence thinking of sustainability issues and coordinate environmental efforts in the nation.

We are also an approved charity and offer tax exemption to donors. SEC continuously engages all sectors of the community by formulating and executing a range of holistic programmes, such as the Singapore Environmental Achievement Awards, Asian Environmental Journalism Awards, School Green Awards, Singapore Green Labelling Scheme, Project: Eco-Office, Project: Eco-Shop and Project: Eco-F&B. In addition, we build a pool of committed volunteers under our Earth Helpers programme. Our Training and Education arm provides the people, public and private sectors with the opportunity to develop awareness, knowledge, skills and tools in order to protect and improve our environment for a sustainable future.

Strong partnerships with corporations, government agencies and other NGOs are valued by us. These partnerships are vital for sustaining our programmes, leading to positive action and change. Over the years, SEC has given strength and direction to the environmental movement in Singapore.

For further information, please visit https://sec.org.sg/.

Philippines

The National Ecolabelling Programme Green Choice Philippines (NELP-GCP) is an ecolabelling programme based on ISO 14024 Guiding Principles and Procedures. It is a voluntary, multiple criteria-based, and third-party programme the aims to encourage clean manufacturing practices and consumption of environmentally preferable products and services. It awards the seal of approval to product or service that meets the environmental criteria established for the product category by a multi-sector Technical Committee. Products with the Green Choice Philippines Seal assures the consumers on its preference for the environment. NELP-GCP is being administered by the Philippine Center for Environmental Protection and Sustainable Development, Inc. (PCEPSDI).

Contact:
Website: https://pcepsdi.org.ph/
E-mail: greenchoicephilippines@pcepsdi.org.ph,
greenchoicephilippines@gmail.com

Denmark, Finland, Norway, Iceland, Sweden

The Nordic Swan Ecolabel
The Nordic Swan Ecolabel is the official Nordic ecolabel supported by all Nordic Governments. It is among the world›s strictest and most recognised environmental certifications.

The Nordic Swan Ecolabel is a Type I environmental labelling program established in 1989 by the Nordic Council of Ministers, connect¬ing policy, people, and businesses with the mission to make it easy to make the environmentally best choice. Nordic Ecolabelling is the non-profit organisation responsible for the Nordic Swan Ecolabel.

The organisation offers independent third-party certification and support for a wide range of product areas and services, ensuring that they comply with the Nordic Swan Ecolabel's strict requirements through documentation and inspections.

30 years of experience and expertise has made the Nordic Swan Ecolabel a powerful tool that paves the way to a sustainable future by giving producers a recipe on how to develop more environmentally sustainable products, and giving consumers credible guidance by helping them identify products that are among the environmentally best.

Globally, you can find more than 25,000 Nordic Swan ecolabelled products. 93% of all Nordic consumers recognise the Nordic Swan Ecolabel as a brand, and 74% believe that the Nordic Swan Ecolabel makes it easier for them to make envi¬ronmentally friendly choices (IPSOS 2019).

Denmark, Finland, Norway, Iceland, Sweden

Securing a sustainable future
The Nordic Swan Ecolabel works to reduce the overall environmental impact from production and consumption and contributes significantly to UN Sustainable Development Goal 12: Responsible consumption and production.

To ensure maximum environmental impact, the Nordic Swan Ecolabel sets product specific requirements and evaluates the environmental impact of a product in all relevant stages of a product lifecycle - from raw materials, production, and use, to waste, re-use and recycling.

Common to all products certified with the Nordic Swan Ecolabel is that they meet strict environmental and health requirements. All requirements must be documented and are verified by Nordic Ecolabelling. Nordic Ecolabelling regularly reviews and tightens the requirements.

Therefore, certifications are time-limited and companies must re-apply to ensure sustainable development.

International website:
Nordic-ecolabel.org
National websites:
Denmark: ecolabel.dk
Sweden: svanen.se
Norway: svanemerket.no (in Norwegian)
Finland: joutsenmerkki.fi (in Finnish)
Iceland: svanurinn.is (in Icelandic)

USA

The Carbonfree® Product Certification is a meaningful, transparent way for you to provide environmentally-responsible, carbon neutral products to your customers. By determining a product's carbon footprint, reducing it where possible and offsetting remaining emissions through our third-party validated carbon reduction projects, companies can:
- Differentiate their brand and product
- Increase sales and market share
- Improve customer loyalty
- Strengthen corporate social responsibility & environmental goals

The Carbonfree® Product Certification Program is proud to be part of Amazon's Climate Pledge Friendly Program!
Carbonfund.org is leading the fight against climate change, making it easy and affordable to reduce & offset climate impact and hasten the transition to a clean energy future.

Contact:

O: 240.247.0630 ext 633
C: 203.257.7808
M: 853 Main Street, East Aurora, NY, 14052

Taiwan

In order to provide guidance to consumers for the purchase of products with high energy efficiency three principal policies have been employed in the promotion of energy efficiency management for energy-consuming equipment and apparatuses in Taiwan; those include Minimum Energy Performance Standard (MEPS), voluntary energy efficiency labeling program and mandatory energy efficiency rating labeling program. At present, Taiwan has announced MEPS requirements for 29 product categories; and 51 product categories for the voluntary energy efficiency labeling program; 17categories of products for the mandatory Energy Efficiency Rating Labeling system.

The related information is on the website www.energylabel.org.tw.

Thailand

The Thai Green Label Scheme was initiated by the Thailand Business Council for Sustainable Development (TBCSD) in October 1993. It was formally launched in August ١٩٩٤ by The Thailand Environment Institute (TEI) and Thai Industrial Standards Institute (TISI). The Green Label is an environmental certification logo awarded to specific products which have less detrimental impact on the environment in comparison with other products serving the same function. The Thai Green Label Scheme applies to all products and services, but not foods, beverage, and pharmaceuticals. Products or services which meet the Thai Green Label criteria may carry the Thai Green Label. Participation in the scheme is voluntary.

Thailand Environment Institute (TEI)
16/151 Muang Thong Thani, Bond Street,
Bangpood, Pakkred, Nonthaburi 11120 THAILAND
Tel. +66 2 503 3333 ext. 303, 315, 116
Fax. +66 2 504 4826-8
Website: http://www.tei.or.th/greenlabel/
Email: lunchakorn@tei.or.th

EUROPE

Established in 1992 and recognized across Europe and worldwide, the EU Ecolabel is a label of environmental excellence that is awarded to products and services meeting high environmental standards throughout their life-cycle: from raw material extraction, to production, distribution and disposal. The EU Ecolabel promotes the circular economy by encouraging producers to generate less waste and CO_2 during the manufacturing process. The EU Ecolabel criteria also encourages companies to develop products that are durable, easy to repair and recycle.

The EU Ecolabel criteria provide exigent guidelines for companies looking to lower their environmental impact and guarantee the efficiency of their environmental actions through third party controls. Furthermore, many companies turn to the EU Ecolabel criteria for guidance on eco-friendly best practices when developing their product lines. The EU Ecolabel helps you identify products and services that have a reduced environmental impact throughout their life cycle, from the extraction of raw material through to production, use and disposal. Recognised throughout Europe, EU Ecolabel is a voluntary label promoting environmental excellence which can be trusted.

Spain , Germany, Italy, Sweden, Greece, Portugal, Poland, Belgium, Netherlands, Estonia, Finland, Austria, Lithuania, Czech Republic, Norway, Cyprus, Ireland, Slovenia, Hungary, Romania, Croatia, Bulgaria, Malta, Slovak Republic, Latvia, Luxembourg, Iceland

Contact and more information via: http://ec.europe.eu

Sweden

TCO Certified is the world-leading sustainability certification for IT products. Covering 11 product categories including computers, mobile devices, display products, and data center products, its comprehensive criteria are designed to drive social and environmental responsibility throughout the product life cycle. Independent verification of criteria compliance is always included. Independent verifiers spend around 20,000 hours every year on tests and audits. Currently, more than 3,500 products from 27 well-known IT brands are certified. The purpose of TCO Certified is to drive progress toward a future where all IT products have a sustainable life cycle, something that requires a collective effort from IT buyers as well as industry. TCO Certified helps the IT industry structure their work with sustainability and offers a platform for continuous improvement. Organizations that buy IT products use the certification as a tool for making more responsible IT product choices.

Contact:
TCO Development |
Linnégatan 14 | 11447 Stockholm, Sweden |
Mobile: +46 (0) 706 358351|
Email: Marketing@tcodevelopment.com

For all People who wish to take care of Climate Change

For all Schools, Libraries, Homes and/or Offices

Available in more than 39,000 booksellers worldwide, including Amazon, Barnes & Noble, Google Play Books, Walmart, ... and many more.

International Environmental Labelling
Set Box Book series (Vol.1-11)
+ **Free** Knowledge Test

Special Thanks to:
United Nations Environment Programme
(UNEP)
& more than 57 Countries

Order Now
online

http://TopTenAward.Net

CHAPTER 5

Type II Environmental Labelling

Type II environmental labelling refers to the claims made on product labels in connection with business centers. This includes familiar claims such as recyclable, ozone-free, 60% phosphate-free, and the like. This type of labelling can be in the form of a mark or sentence on the product packaging. Some of them are valid environmental claims—and some can be completely misleading.

Usually, all countries have laws against deceptive advertisements, so why has the International Organization for Standardization discussed this issue? The answer is that it is not clear whether the environmental claims have a technical basis or whether the ad is meaningless.

Most countries have guidelines at the national level to help producers and consumers know what constitutes a true, scientifically valid claim.
There is a national standard on this in Canada. In Australia, the Consumer Commission has published guidance on this, and there are similar examples in other countries.

Canada

Energy Moon Ecolabel established in Coquitlam, British Columbia, Canada in 2021. ENERGY MOON is an international ecolabel focused on Energy Saving in different industries and categorized as Type II Environmental Labelling. It's defined as 'self-declared' energy saving claims made by manufacturers and businesses based on ISO 14020 series of standards, the claimant can declare the Energy Saving objectives and targets and also propose programmes for achiving the defined objectives. However, this declaration will be verifiable.

Energy Moon
Coquitlam, BC CANADA

Email: info@energymoon.org
Web: www.energymoon.org

Canada

Environmental Sustain for Future kids established in Vancouver, BC Canada in 2020. (ESFK) is an international ecolabel focused on taking care of environment for future of kids.

ESFK defined as 'self-declared' environmental claims made by manufacturers and businesses based on ISO 14020 series of standards, the claimant can declare the environmental objectives and targets in relation to taking care of environment for future kids. However, this declaration will be verifiable.

Environmental Sustain for Future Kids
Vancouver, BC CANADA

Email: info@esfk.org
Web: www.esfk.org

Five Ways To Saving Energy by Kids:

1. Pick Up a Book instead of turning on TV and/or Computer
2. Keep the Doors Closed
3. Use the Sun to Dry Wet Clothes
4. Utilize Natural Light
5. Turning off Switches

CHAPTER 6

Type III Environmental Labelling

Type III environmental labelling is a distinct form of third-party environmental labelling pattern designed to avoid the difficulties that can result from type I labelling. Technical committee for Environment of International organization for Standardization has undertaken a new project to standardize guidelines and Type III labelling methods. One of the main objections raised by industries to Type I labelling is the basis for its management.

Due to the nature of the system, less than 50% of the various products on the market can meet the criteria and qualify for Type I Labelling. As long as the industry is the main supporter of other third-party models for quality systems, it is sometimes difficult for an industry to support a program that can only benefit 15% of its members. This type of labelling is currently practiced in some countries, such as Sweden, Canada, and the United States. Choosing the right product has never been easy, but Type III labelling will help because each product can have a label that describes its environmental performance and is certified by a third-party company. Consumers can then compare labels and choose their favorite products.

中華民國 能源效率標示

每年耗電量 約 XXX 度

本產品能源效率為第 1 級

名 稱	無風管空氣調節機
型 號	00-000000
額 定 冷氣能力	X.X kW
CSPF 冷氣季節 性能因數	X.X kWh/kWh

本產品能源效率符合國家標準，其分級係依經濟部 104 年 8 月 11 日經能字第 10404603780 號公告之能源效率分級基準表標示

登錄編號：

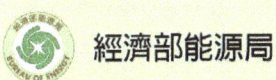

經濟部能源局

5 — 用電較多

4

3

2

1 級 — 1 — 用電較少

Labels serve the purpose of allowing consumers to make comparisons and informed choices from among products and or services in a category. Environmental labels focus primarily on consumption rather than production of goods.

Ecolabels communicate the environmental impacts over the life cycle of the product "from cradle to grave".

CHAPTER 7

What is Renewable energy ?

Renewable energy is energy that has been derived from earth's natural resources that are not finite or exhaustible, such as wind and sunlight. Renewable energy is an alternative to the traditional energy that relies on fossil fuels, and it tends to be much less harmful to the environment.

Types of Renewable Energy :

Solar energy is derived by capturing radiant energy from sunlight and converting it into heat, electricity, or hot water. Photovoltaic (PV) systems can convert direct sunlight into electricity through the use of solar cells.

Wind farms capture the energy of wind flow by using turbines and converting it into electricity. There are several forms of systems used to convert wind energy and each vary. Commercial grade wind-powered generating systems can power many different organizations, while single-wind turbines are used to help supplement pre-existing energy organizations.

Hydroelectric, Dams are what people most associate when it comes to hydroelectric power. Water flows through the dam's turbines to produce electricity, known as pumped-storage hydropower. Run-of-river hydropower uses a channel to funnel water through rather than powering it through a dam.

Geothermal heat is heat that is trapped beneath the earth's crust from the formation of the Earth 4.5 billion years ago and from radioactive decay. Sometimes large amounts of this heat escapes naturally, but all at once, resulting in familiar occurrences, such as volcanic eruptions and geysers.

Ocean, the ocean can produce two types of energy: thermal and mechanical. Ocean thermal energy relies on warm water surface temperatures to generate energy through a variety of different systems. Ocean mechanical energy uses the ebbs and flows of the tides to generate energy, which is created by the earth's rotation and gravity from the moon.

Hydrogen needs to be combined with other elements, such as oxygen to make water as it does not occur naturally as a gas on its own. When hydrogen is separated from another element it can be used for both fuel and electricity.

Bioenergy is a renewable energy derived from biomass. Biomass is organic matter that comes from recently living plants and organisms. Using wood in your fireplace is an example of biomass that most people are familiar with.

Renewable Energy: What Can You Do?

As a consumer you have several opportunities to make an impact on improving the environment through the choice of a greener energy solution. If you're a homeowner, you have the option of installing solar panels in your home. Solar panels not only reduce your energy costs, but help improve your standard of living with a safer, more eco-friendlier energy choice that doesn't depend on resources that harm the environment. There are also alternatives for a greener way of life offered by your electric companies. Just Energy allows consumers to choose green energy options that help you reduce your footprint with energy offsets.

Solar

Solar energy is derived by capturing radiant energy from sunlight and converting it into heat, electricity, or hot water. Photovoltaic (PV) systems can convert direct sunlight into electricity through the use of solar cells.

Benefits

One of the benefits of solar energy is that sunlight is functionally endless. With the technology to harvest it, there is a limitless supply of solar energy, meaning it could render fossil fuels obsolete. Relying on solar energy rather than fossil fuels also helps us improve public health and environmental conditions. In the long term, solar energy could also eliminate energy costs, and in the short term, reduce your energy bills. Many federal local, state, and federal governments also incentivize the investment in solar energy by providing rebates or tax credits.

Current Limitations

Although solar energy will save you money in the long run, it tends to be a significant upfront cost and is an unrealistic expenses for most households. For personal homes, homeowners also need to have the ample sunlight and space to arrange their solar panels, which limits who can realistically adopt this technology at the individual level.

Wind

Wind farms capture the energy of wind flow by using turbines and converting it into electricity. There are several forms of systems used to convert wind energy and each vary. Commercial grade wind-powered generating systems can power many different organizations, while single-wind turbines are used to help supplement pre-existing energy organizations. Another form is utility-scale wind farms, which are purchased by contract or wholesale. Technically, wind energy is a form of solar energy. The phenomenon we call "wind" is caused by the differences in temperature in the atmosphere combined with the rotation of Earth and the geography of the planet.

Benefits

Wind energy is a clean energy source, which means that it doesn't pollute the air like other forms of energy. Wind energy doesn't produce carbon dioxide, or release any harmful products that can cause environmental degradation or negatively affect human health like smog, acid rain, or other heat-trapping gases. Investment in wind energy technology can also open up new avenues for jobs and job training, as the turbines on farms need to be serviced and maintained to keep running.

Current Limitations

Since wind farms tend to be built in rural or remote areas, they are usually far from bustling cities where the electricity is needed most. Wind energy must be transported via transition lines, leading to higher costs. Although wind turbines produce very little pollution, some cities oppose them since they dominate skylines and generate noise. Wind turbines also threaten local wildlife like birds, which are sometimes killed by striking the arms of the turbine while flying.

Hydroelectric

Dams are what people most associate when it comes to hydroelectric power. Water flows through the dam's turbines to produce electricity, known as pumped-storage hydropower. Run-of-river hydropower uses a channel to funnel water through rather than powering it through a dam.

Benefits

Hydroelectric power is very versatile and can be generated using both large scale projects, like the Hoover Dam, and small scale projects like underwater turbines and lower dams on small rivers and streams. Hydroelectric power does not generate pollution, and therefore is a much more environmentally-friendly energy option for our environment.

Current Limitations

Most hydroelectricity facilities use more energy than they are able to produce for consumption. The storage systems may need to use fossil fuel to pump water. Although hydroelectric power does not pollute the air, it disrupts waterways and negatively affects the animals that live in them, changing water levels, currents, and migration paths for many fish and other freshwater ecosystems.

Geothermal

Geothermal heat is heat that is trapped beneath the earth's crust from the formation of the Earth 4.5 billion years ago and from radioactive decay. Sometimes large amounts of this heat escapes naturally, but all at once, resulting in familiar occurrences, such as volcanic eruptions and geysers. This heat can be captured and used to produce geothermal energy by using steam that comes from the heated water pumping below the surface, which then rises to the top and can be used to operate a turbine.

Benefits

Geothermal energy is not as common as other types of renewable energy sources, but it has a significant potential for energy supply. Since it can be built underground, it leaves very little footprint on land. Geothermal energy is naturally replenished and therefore does not run a risk of depleting (on a human timescale).

Current Limitations

Cost plays a major factor when it comes to disadvantages of geothermal energy. Not only is it costly to build the infrastructure, but another major concern is its vulnerability to earthquakes in certain regions of the world.

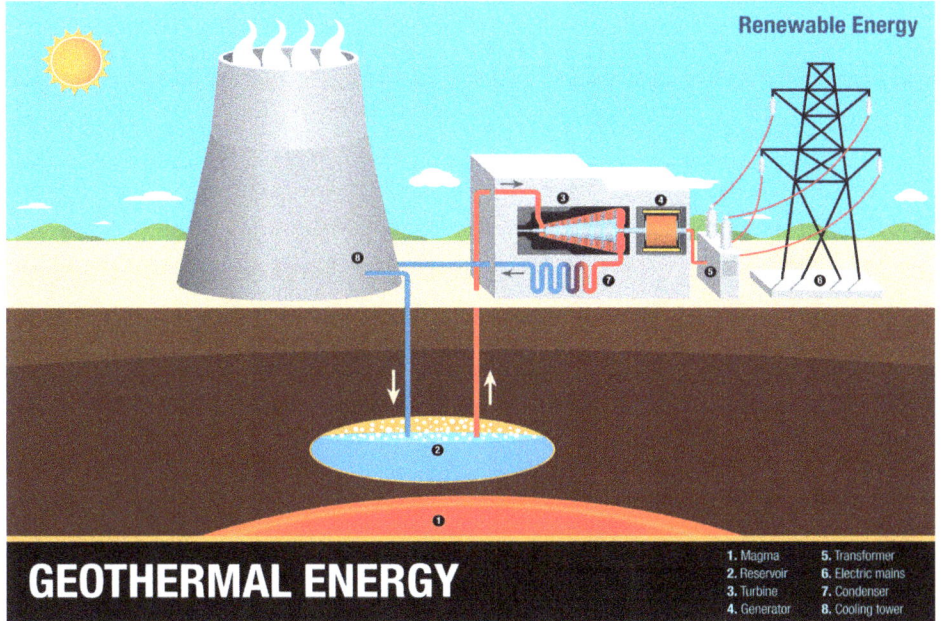

Ocean

The ocean can produce two types of energy: thermal and mechanical. Ocean thermal energy relies on warm water surface temperatures to generate energy through a variety of different systems. Ocean mechanical energy uses the ebbs and flows of the tides to generate energy, which is created by the earth's rotation and gravity from the moon.

Benefits

Unlike other forms of renewable energy, wave energy is predictable and it's easy to estimate the amount of energy that will be produced. Instead of relying on varying factors, such as sun and wind, wave energy is much more consistent. This type of renewable energy is also abundant, the most populated cities tend to be near oceans and harbors, making it easier to harness this energy for the local population. The potential of wave energy is an astounding as yet untapped energy resource with an estimated ability to produce 2640 TWh/yr. Just 1 TWh/yr of energy can power around 93,850 average U.S. homes with power annually, or about twice than the number of homes that currently exist in the U.S. at present.

Current Limitations

Those who live near the ocean definitely benefit from wave energy, but those who live in landlocked states won't have ready access to this energy. Another disadvantage to ocean energy is that it can disturb the ocean's many delicate ecosystems. Although it is a very clean source of energy, large machinery needs to be built nearby to help capture this form energy, which can cause disruptions to the ocean floor and the sea life that habitats it. Another factor to consider is weather, when rough weather occurs it changes the consistency of the waves, thus producing lower energy output when compared to normal waves without stormy weather.

Hydrogen

Hydrogen needs to be combined with other elements, such as oxygen to make water as it does not occur naturally as a gas on its own. When hydrogen is separated from another element it can be used for both fuel and electricity.

Benefits

Hydrogen can be used as a clean burning fuel, which leads to less pollution and a cleaner environment. It can also be used for fuel cells which are similar to batteries and can be used for powering an electric motor.

Current Limitations

Since hydrogen needs energy to be produced, it is inefficient when it comes to preventing pollution.

Biomass

Bioenergy is a renewable energy derived from biomass. Biomass is organic matter that comes from recently living plants and organisms. Using wood in your fireplace is an example of biomass that most people are familiar with.

There are various methods used to generate energy through the use of biomass. This can be done by burning biomass, or harnessing methane gas which is produced by the natural decomposition of organic materials in ponds or even landfills.

Benefits

The use of biomass in energy production creates carbon dioxide that is put into the air, but the regeneration of plants consumes the same amount of carbon dioxide, which is said to create a balanced atmosphere. Biomass can be used in a number of different ways in our daily lives, not only for personal use, but businesses as well. In 2017, energy from biomass made up about 5% of the total energy used in the U.S. This energy came from wood, biofuels like ethanol, and energy generated from methane captured from landfills or by burning municipal waste.

Current Limitations

Although new plants need carbon dioxide to grow, plants take time to grow. We also don't yet have widespread technology that can use biomass in lieu of fossil fuels.

CHAPTER 8

The Future of Renewable energy

Take a moment to close your eyes and imagine the world ten, twenty, fifty years from now. How do you heat your home? What do our energy systems look like? How about our cars – how are they fueled?

In an ideal world, renewable energy will become the primary source of the planet's energy, as opposed to traditional energy sources, like fossil fuels (which release harmful carbon emissions and pollution into the atmosphere). So, what does the future of renewable energy actually look like? Time will tell – but these crazy, cool, new innovations may provide a glimpse into the future of renewables:

Solar Powered Panels that Chase the Sun

Genius in its simplicity, this new technology overcomes one of the biggest challenges facing solar power – clouds and inclement weather. These solar panels actually reposition themselves to soak in the most possible sunlight, resulting in much higher levels of efficiency.

Solar/Wind Hybrids

As solar and wind technologies continue to improve, scientists and engineers are experimenting with ways to make both more efficient. Bring on the superhero of renewable energy: solar and wind hybrids. This technology combines wind turbines with solar photovoltaic (PV) panels to produce higher levels of energy – and studies have found that they are nearly twice as efficient.

Energy From Unusual Sources

You've heard about energy from the wind, from the sun, and even from compost or other organic sources, but how about algae? It's true – "algae energy" is a concept that scientists are currently developing. Another awesome technology: these batteries made from wood (oh hey there, bioenergy). Color us impressed.

Do-It-Yourself Renewable Energy

We dream of a world with solar panels on every roof, wind turbines in every backyard. Is this a realistic dream? Scientists and engineers are getting closer every day. Even today, some dedicated homeowners have taken pains to install their own personal systems of solar power to heat/power their homes – a trend we hope to see continue well into the future.

Nuclear Energy Is Extraordinary

Nuclear energy comes from splitting atoms in a reactor to heat water into steam, turn a turbine and generate electricity. Ninety-four nuclear reactors in 28 states generate nearly 20 percent of the nation's electricity, all without carbon emissions because reactors use uranium, not fossil fuels. These plants are always on: well-operated to avoid interruptions and built to withstand extreme weather, supporting the grid 24/7.

FIVE FREE & EASY
Ways To Save Energy in your home:

1. Turn off the fan when you leave a room. ...
2. Close your drapes or drop your window shades during the day. ...
3. Wash your clothes in cold water. ...
4. Wrap or cover foods and drinks in the refrigerator. ...
5. Always use the cold water faucet, unless you really want hot water.

CHAPTER 9

Top Ten Award International Network Environmental Pioneers

Top Ten Award international Network (TTAIN) was established in 2012 to recognize outstanding individuals, groups, companies, organizations representing the best in the public works profession. TTAIN publishing books related to international Eco-labeling plans to increase public knowledge in purchasing based on the environmental impacts of products. We introduce in each volume some of the organizations that are doing their best in relation to taking care of the environmnet.

世界省エネルギー等ビジネス推進協議会
Japanese Business Alliance for Smart Energy Worldwide

Japan

Japanese Business Alliance for Smart Energy Worldwide is a coalition of Japanese companies which have a variety of excellent energy technologies and products and wish to make significant contribution to the global environment and decarbonization by spreading their advanced energy efficient technologies and products to the world. Currently more than 40 Japanese companies of broad business areas are becoming its members and remarkably serving for the global energy conservation and sustainable development. JASE-W is aiming to disseminate these technologies especially to the emerging countries, with particular focus on, for example, Zero Energy Building, Combined Heat and Power, Waste to Energy, etc.

contact:
 jase-w@eccj.or.jp

UNEP

The United Nations Environment Programme (UNEP) is the leading global environmental authority that sets the global environmental agenda, promotes the coherent implementation of the environmental dimension of sustainable development within the United Nations system, and serves as an authoritative advocate for the global environment.

Our mission is to provide leadership and encourage partnership in caring for the environment by inspiring, informing, and enabling nations and peoples to improve their quality of life without compromising that of future generations.

Headquartered in Nairobi, Kenya, we work through our divisions as well as our regional, liaison and out-posted offices and a growing network of collaborating centres of excellence. We also host several environmental conventions, secretariats and inter-agency coordinating bodies. UN Environment is led by our Executive Director.

We categorize our work into seven broad thematic areas: climate change, disasters and conflicts, ecosystem management, environmental governance, chemicals and waste, resource efficiency, and environment under review. In all of our work, we maintain our overarching commitment to sustainability.

Website: www.unep.org

Bibliography

Bibliography:

Andrews, R.N.L. 1998. Environmental regulation and business 'self-regulation'. Policy Sciences 31(3): 177-197.

Apodaca, Julia, "Market Potential of Organically Grown Cotton as a Niche Crop." Natural Fibers Research and Information Center, Bureau of Business Research, University of Texas at Austin, Paper presented at the Beltwide Cotton Conference in Nashville, TN, January 1992.

Asadi, J., "International Environmental Labelling, Economic Consequencies, Export Magazine, July 2001

Asadi, J. 2008. Mobile Phone as management systems tools, ISO Magazine, Vol.8, No.1

Asadi, J., Eco-Labelling Standards, National Standard Magazine, Sep. 2004.

Assocs., Cambridge MA and G. Davis, U. Tenn, Knoxville,TN. (68-W6-0021): xiii+76+226pp.

Balter, M. 1999. Scientific cross-claims fly in continuing beef war. Science (May 28) 284: 1453-1455.

Belsley, D.A., Kuh, E., and Welsch, R.E. (1980), Regression Diagnostics, New York: John Wiley & Sons, Inc.

Birett, M. J. 1997. Encouraging Green Procurement Practices in Business: A Canadian Case Study in Program Development (108-118). in Greener Purchasing : Opportunities and Innovation. Sheffield, Greenleaf Publishing 325p.

Bowen, Nicola, World Agrochemical Markets, PJB Publications Ltd., March 1991.

Bureau of Ocean Energy Management, Ocean Wave Energy, Retrieved From: https://www.boem.gov/Ocean-Wave-Energy/

Burnside, A., (1990), Keen on Green, Marketing, 17 May, pp35-36

Butler, D., (1990), A Deeper Shade of Green, Management Today, June, pp74-79

Cairncross, F. 1995. Green, Inc.: A guide to business and the environment. London, Earthscan. 277p.

Cason, T. N. and L. Gangadharan, (2002), Environmental Labeling and Charter, M. (ed.) 1992. Greener marketing: a responsible approach to business. Sheffield, Greenleaf Publishing 403p.

Chemical Week, 1999. Europe's Beef Ban Tests Precautionary Principle. (August 11).

CHOI, J.P. Brand Extension as Informational Leverage. Review of Eco- nomic Studies, Vol. 65 (1998), pp. 655-669.

Conway, G. 2000. Genetically modified crops: risks and promise.

Corrado, M., (1989), The Greening Consumer in Britain, MORI, London

Corrado, M., (1997), Green Behaviour – Sustainable Trends, Sustainable Lives?, MORI, london, accessed via countries. Manila, Asian Development Bank 33p.

Cropper, M.L., L.D. Deck, and K.E. McConnell. "On the choice of Functional Forms for Hedonic Price Functions," Review of Economics and Statistics 70(1988): 668-675.

Darbi, M. R. and E. Karni, (1973), Free Competition and the Optimal

Davis, G. 1998. Environmental Labeling Issues, Policies, and Practices Worldwide. Washington, DC. EPA, 216p.

Dawkins, K. 1996. Eco-labeling: consumer's right-to-know or restrictive business practice? Minneapolis, Minn., Institute for Agriculture and Trade Policy.

Di Leva, C. E. 1998. International Environmental Law and Development. Georgetown Interna. Environ. Law Review 10 (2): 502-549.

Economics and Management 43, 339-359.

Eiderstroem, E. 1997. Eco-labeling: Swedish Style. Forum for Applied Research in Public Policy 141(4).

Elkington, J. and Hailes, J. 1990. The green consumer guide: You can buy products that don't cost the earth. New York, Viking Press. 96p.

EMONS, W. Credence Goods and Fraudulent Experts. RAND Journal of Economics, Vol. 28 (1997), pp. 107-119.

EMONS, W. Credence Goods Monopolists. International Journal of In- dustrial Organization, Vol. 19 (2001), pp. 375-389.

Energy.gov, Advantages and Challenges of Wind Energy, Retrieved from: https://www.energy.gov/eere/wind/advantages-and-challenges-wind-energy

Energy.gov, Advantages and Challenges of Wind Energy, Retrieved from: https://www.energy.gov/eere/wind/advantages-and-challenges-wind-energy

Environment Canada 1997. Towards Greener Government Procurement: An Environment Canada Case Study (pp. 31-46). in Greener Purchasing: Opportunities and Innovations.

Environmental Protection Agency 742-R-98-009, (1998),

Environmentalist 17 (2): 125-133.

Erskine, C.C. and Collins, L. 1996. Eco-labeling in the EU: a comparative study of the pulp and paper industry in the UK and Sweden. European Environment 17 (2) : 40-47.

Erskine, C.C. and Collins, L. 1997. "Eco-labeling: Success or failure?".

Ethical Consumer, (1995), Co-op Supermarkets take up Ethics, EC36, June/July, p4

Ethical Consumer, (June 1996), Green Cons, EC41, June, p5

European Communities, Commission of the, 1996. Eco-label revision.

European Communities, Commission of the. 1996. Conservation of West Africa's forests through certification. UN Courier 157: 71-73.

European Union official website: https://ec.europa.eu/info/about-european-commission/contact_en

Feenstra, R.C. "Exact Hedonic Price Indexes," Review of Economics and Statistics 77 (1995): 634-653.

Feenstra, R.C., and J.A. Levinsohn. "Estimating Markups and Market Conduct with Multidimensional Product Attributes," Review of Economic Studies (62 (1995): 19-52.

Forest Stewardship Council: "Principles and criteria for forest stewardship" Document 1.2: <http://www.fscoax.org>

Forsyth, K. 1999. Will consumers pay more for certified wood products? Journal of Forestry 97 (2) : 18-22.

Freeman, A. M III. The Measurement of Environmental and Resource Values. Theory and Methods. Washington D.C.: Resource for the Future, 1993.

Friends of the Earth, 1993. Timber certification and eco-labeling. London, FOE:

Graves, P., J.C. Murdoch, M.A. Thayer, and D. Waldman. "The Robustness of Hedonic Price Estimation: Urban Air Quality," Land Economics 64(1988): 220-233.

Halvorsen, R. and R. Palmquist. "The Interpretation of Dummy Variables in Semilogarithmic Equations." American Economic Review 70:474-75 (1980).

Imhoff, Dan, and Grose, Lynda, and Carra, Roberto., "Organic Cotton Exhibit," Mimeo. Simple Life and distributed the Texas Organic Cotton Marketing Cooperative, O'Donnell, Texas (1996).

Imhoff, Dan. "Growing Pains: Organic Cotton Tests the Fiber of Growers and Manufacturers Alike," reprinted on Simple Life's web page (simplelife.com), but first printed by Farmer to Farmer, December 1995.

Incomplete Consumer Information in Laboratory Markets. Journal of Environmental labeling.

ISO 14020, ISO 14021,ISO 14024,ISO 14025, International Organization for Standardization.

Kennedy, P.E. "Estimation with Correctly Interpreted Dummy Variables in Semilogarithmic Equations," American Economic Review 71: 801 (1981).

Kirchho®, S., (2000), Green Business and Blue Angels.

Kraus, Jeff. Lab Technician at the North Carolina School of Textiles.

Labeling Issues, Policies and Practices Worldwide.

Lamport, L. 1998. The cast of (timber) certifiers: who are they? International J. Ecoforestry 11(4): 118-122.

Large Scale impoverishment of Amazonian forests by logging and fire. 1999.

Lathrop, K.W. and Centner, T.J. 1998. Eco-labeling and ISO 14000: An analysis of US regulatory systems and issues concerning adoption of type II standards. Environmental

Lee, J. et al. 1996. Trade related environmental measures; sizing and comparing impacts.

Lehtonen, Markku. 1997. Criteria in Environmental Labeling: A comparative Analysis on Environmental Criteria in Selected Labeling Schemes. Geneva, UNEP. 148p.

LIEBI, T. Trusting Labels: A Matter of Numbers? Working Paper Uni versity of Bern, No. 0201 (2002).

Lindstrom, T. 1999. Forest Certification: The View from Europe's NIPFs. Journal of Forestry 97(3): 25-31. London

Losey, J.E., Rayor, L.S. & Carter, M.E. 1999. Transgenic pollen harms monarch larvae. Nature 399 20 May): p.214.

Management 22 (2) : 163-172.

Mattoo, A. and H. V. Singh, (1994), Eco-Labelling: Policy Considera-

Michaels, R. G., and V. K. Smith. "Market Segmentation And Valuing Amenities With Hedonic Models: The Case Of Hazardous Waste Sites," Journal of Urban Economics, 1990 28(2), 223-242.

Mintel, (1991), The Green Consumer I, May

Mintel, (1994), The Green Consumer, Mintel Special Report

Moraga-Gonzalez, J. L. and N. Padrón-Fumero, (2002),

NCC, (1996a), Green Claims – a consumer investigation into marketing claims about the environment,

NCC, (1996b), Shades of Green – consumers' attitudes to green shopping, National Consumer Council,

Nelson , P."Information and Consumer Behaviour," Journal of Political Economy 78 (1970): 311-329..

Nicholson-Lord, D., (1993) 'Tis the Season to be Green, The Independent, 20 December

Nuttall, N., (1993), Shoppers can cross green products off their lists, The Times, 3 July

OCDE/GD(97)105. Paris, OECD. 81p.

OECD. "Ec-labelling: Actual Effects of Selected Programmes," OCDE/GD (97) 105, 1997, Paris. (available on line at http://www.oecd.org/env/eco/books.htm#tradermono)

OECD. 1997a. Case study on eco-labeling schemes. Paris, OECD (30 Dec):

OECD. 1997b. Eco-labeling: Actual Effects of Selected Programs.

Osborne, L. "Market Structure, Hedonic Models, and the Valuation of Environmental Amenities." Unpublished Ph.D. dissertation. North Carolina State University, 1995.

Osborne, L., and V. K. Smith. "Environmental Amenities, Product Differentiation, and market Power," Mimeo, 1997.

Ozanne, L.K. and Vlosky, R.P. 1996. Wood products environmental certification: the United States perspective". Forestry Chronicle 72 (2) : 157-165.

Palmquist, R. B., F. M. Roka, and T.Vukina. "Hog Operations, Environmental Effects, and Residential Property Values," Land Economics 73(1), (1997): 114-24.

Palmquist, R.B. "Hedonic Methods," in J.B Braden and C.D. Kolstad, eds. Measuring the Demand for Environmental Improvement. Amsterdam, NL: Elsevier, 1991.

Pento, T. 1997. Implementation of Public Green Procurement Programs (22-31) in Greener Purchasing: Opportunities and Innovations. Sheffield, Greenleaf Publ. 325 p.

Perloff, J. "Industrial Organization Lecture Notes," Mimeo. University of California at Berkeley (1985).

Plant, C. and Plant, J. 1991. Green business: hope or hoax? Philadelphia, New Society Publishers 136 p.

Polak, J. and Bergholm, K. 1997. Eco-labeling and trade: a cooperative approach (Jan.): Policy in a Green Market. Environmental and Resource Economics 22, 419-

Poore, M.E.D. et al. 1989. No timber without trees. London, Earthscan. 352p.

Raff, D. M.G., and M. Trajtenberg. "Quality-Adjusted Prices for the American Automobile Industry: 1906-1940." NBER Working Paper Series, Working Paper No. 5035, February 1995.

Rastogi, J. 1998. What's Behind the Label? Complexities of Certified Wood. Ecoforestry 13 (2): 38-42.

Roberts, J. T. 1998. Emerging global environment standards: prospects and perils. Journal of Developing Societies 14 (1): 144-163.

Rosen, S., "Hedonic Prices and Implicit Markets: Product Differentiation in Pure Competition." Journal of Political Economy. 82: 34-55 (1974).

Ross, B. 1997. Eco-friendly procurement training course for UN HCR. : 126 p.

Ryan, S., and Skipworth, M., (1993), Consumers turn their backs on green revolution, The Times, 4 April

Salzman, J. 1997. Informing the Green Consumer: The Debate over the Use and Abuse of Environmental Labels. Journal of Industrial Ecology 1 (2): 11-22.

Sanders, W. 1997. Environmentally Preferable Purchasing: The US Experience (946-960) in Greener Purchasing: Opportunities and Innovations. Sheffield, Greenleaf Publ. 325p.

Sayre, D. 1996. Inside ISO 14000: The competitive advantage of environmental management. Delray Beach FL., St. Lucie Press. 232p.

SHAPIRO, C. Premiums for High Quality Products as Returns to Reputa- tion. Quarterly Journal of Economics, Vol. 98, No. 4 (1983), pp. 659-680.

Stillwell, M. and van Dyke, B. 1999. An activists handbook on genetically modified organisms and the WTO. Washington DC., The Consumer's Choice Council: 20 p.

Teisl, M. F., B. Roe, and R. L. Hicks. "Can Eco-labels tune a market? Evidence from dolphin-safe labeling," Presented paper at the 1997 American Agricultural Economics Association Meetings, Toronto.

THE GERSEN, C. Psychological Determinants of Paying Attention to Eco- Labels in Purchase Decisions: Model Development and Multinational Vali- dation. Journal of Consumer Policy, Vol. 23, No. 4 (2000), pp. 285-313.

Tibor, T. and Feldman, I. 1995. ISO 14000: a guide to the new environmental management standards. Burr Ridge Ill., Irwin Professional Publ. 250 p.

Torre, I. de la, & Batker, D. K. (n.d.) 1999-2000. Prawn to trade: prawn to consume. Graham WA., Industrial Shrimp Action Network (isatorre@seanet.com), [and] Asia –Pacific

Townsend, M. 1998. Making things greener: motivations and influences in the greening of manufacturing. Aldershot, England, Ashgate Publisher. 203p.

U.S. Energy Information Administration, What is U.S. Electricity Generation by Energy Source?, Retrieved From: https://www.eia.gov/tools/faqs/faq.php?id=427&t=3

U.S. Energy Information Administration, Biomass Explained, Retrieved From: https://www.eia.gov/energyexplained/?page=biomass_home

U.S. Environmental Protection Agency. National Water Quality Fact Inventory: 1990 Report to Congress. EPA 503-9-92-006, Apr. 1992.

UK Eco-labelling Board website, accessed via http://www.ecosite.co.uk/Ecolabel-UK/

US Environmental Protection Agency (EPA742-R-99-001): 40 p. <www.epa.gov/opptintr/epp>

US EPA, 1993. Determinants of effectiveness for environmental certification and labeling programs. Washington, D.C., US Environmental Protect

US EPA, 1993. Status report on the use of environmental labels worldwide. Washington, D.C., US Environmental Protection Agency (742-R-93-001 September).

US EPA, 1993. The use of life-cycle assessment in environmental labeling. Washington, D.C., US Environmental Protection Agency (742-R-93-003 September).

US EPA, 1998. Environmental labeling: issues, policies, and practices worldwide. Washington DC., Environmental Protection Agency, Pollution Prevention Division Prepared by Abt

US EPA, 1999. Comprehensive procurement guidelines (CPG) program. Washington, D.C., US Environmental Protection Agency: <www.epa.gov/cpg>

US EPA, 1999. Environmentally preferable purchasing program: Private sector pioneers: How companies are incorporating environmentally preferable purchases. Washington, D.C.,

USG, 1993. Federal acquisition, recycling, and waste prevention. Washington DC., Executive Order: (20 October).

USG, 1998. Greening the government through waste prevention, recycling, and federal acquisition. Washington, D.C., Executive Order 13101 (September).

Van der Grijp, N. 1998. The Greening of Public Procurement in the Netherlands (60-71) in Greener Purchasing: Opportunities and Innovations. Sheffield, Greenleaf Pub. 325 p.

Vanclay, J.K. 1996. Lessons from the Queensland rainforests: steps towards sustainability. J. Sustainable Forestry 3 (2/3): 1-25.

Vidal, J., (1993), Shopping for a paler shade of green, The Guardian, 7 April

Voluntary Overcompliance. Journal of Economic Behavior and Organization

Von Felbert, D. 1995. Trade, environment and aid. Paris, OECD Observer 195: 6-10.

Ward, H. 1997. Review of European Community and International Environmental Law 6 (2): 139-147.

Wasik, John, F. Green Marketing and Management: a Global Perspective, Blackwell Business: Cambridge, Mass, 1996.

West, K. 1995. Ecolabels: the industrialization of environmental standards. The Ecologist (Jan/Feb) 25: 16-20.

Worcester, R., (1995), Business and the Environment – in the aftermath of Brent Spar and BSE, MORI,

World Commission on Forests and Sustainable Development: Final Report. <http://iisd.ca/wcfsd>.

Zarrilli, S., V. Jha, and R. Vossenaar, eds. Eco-labelling and International Trade, St martin Press, Inc. New-York, 1997.

Appendix I: Search by Logos

Here you can search the logos in this volume. It will help you to better undersand the Ecolabels you may encounter while shopping. Buying Eco-products will aid in having a better environment with minimum polution during production processes. Three important parameteres for shopping are **quality**, **price** & **environmental impacts** of the products. Consumers can access the database by scanning the QR code featured on the new energy labels. The database offers detailed information for all registered products, such as for a fridge, the energy efficiency performances, the model dimensions, the type of compartments and their individual volume, or the minimum guarantee offered by the supplier.

INTERNATIONAL ENVIRONMENTAL LABELLING VOL.2 · 83

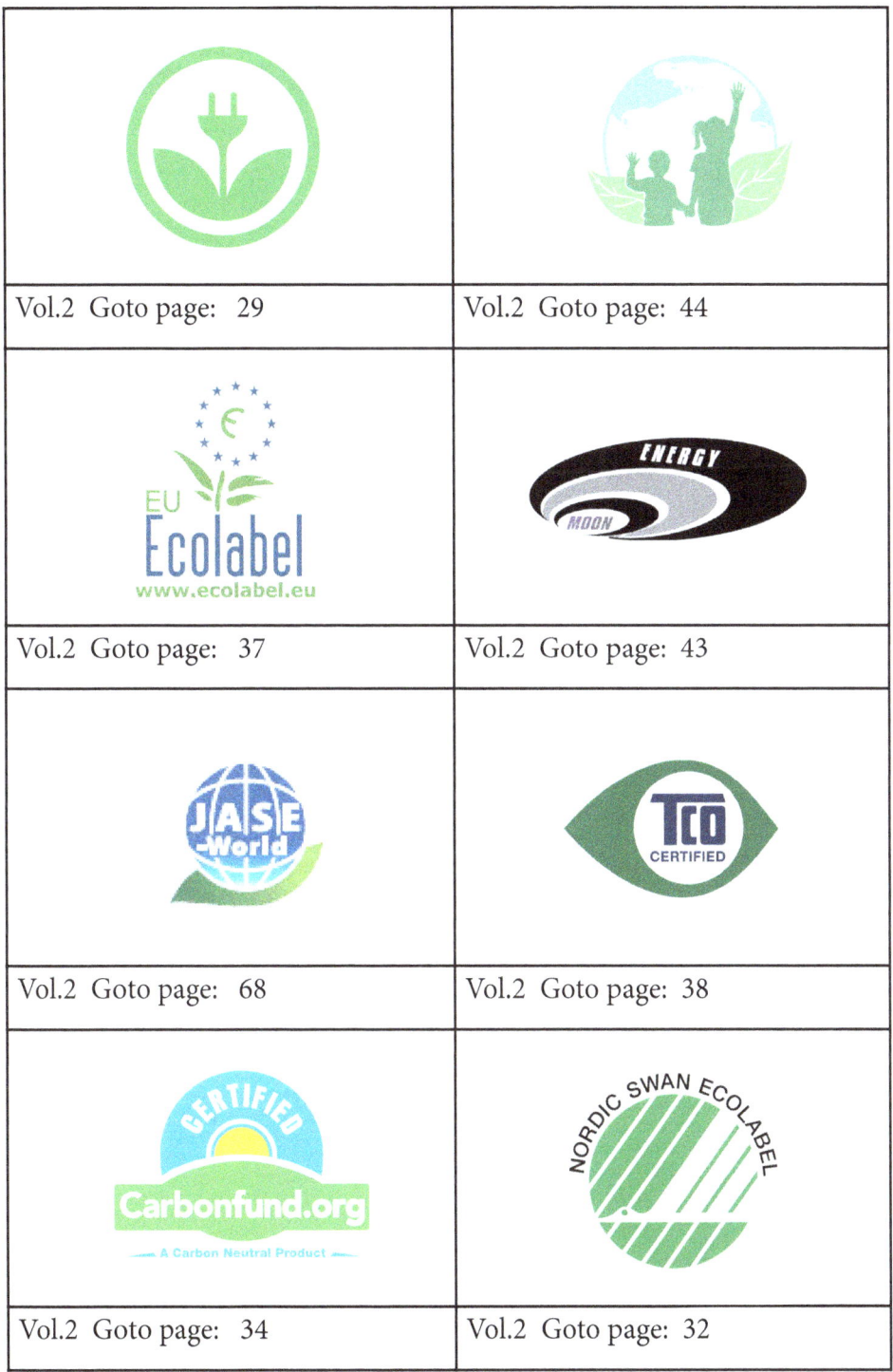

Vol.2 Goto page: 29	Vol.2 Goto page: 44
Vol.2 Goto page: 37	Vol.2 Goto page: 43
Vol.2 Goto page: 68	Vol.2 Goto page: 38
Vol.2 Goto page: 34	Vol.2 Goto page: 32

Vol.2 Goto page: 43	Vol.2 Goto page: 35
Vol.2 Goto page: 44	Vol.2 Goto page: 31
Vol.2 Goto page: 47	Vol.2 Goto page: 30
Vol.2 Goto page: 33	Vol.2 Goto page: 36

INTERNATIONAL ENVIRONMENTAL LABELLING VOL.2 · 85

Vol.2 Goto page: 97	Vol.1 Goto page: 98
Vol.2 Goto page: 86	Vol.2 Goto page: 89
Vol.2 Goto page: 91	Vol.2 Goto page: 94
Vol.2 Goto page: 92	Vol.2 Goto page: 93

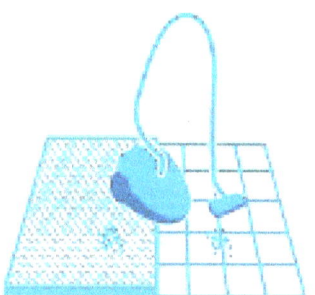

Vacuum Cleaner

Ecodesign requirements

- Energy efficiency
- Performance
- Product information

Standby and off mode

A wide range of electric devices can have standby and off modes.

These are required to switch into a low power mode (such as standby) after a reasonable amount of time and they must not consume more than 0.5 Watts in standby or in off mode.

Energy savings

By switching to one of the most energy efficient vacuum cleaners, you can save $100 over the lifetime of the product. With more efficient vacuum cleaners, Europe as a whole can save up to 20 TWh of electricity per year by 2020. This is equivalent to the annual household electricity consumption of Belgium. It also means over 6 million tonnes of CO_2 will not be emitted – about the annual emissions of eight medium-sized power plants.

Floor polishers, robot vacuums, mattress cleaners, and hand-held and battery operated vacuum cleaners are excluded from these regulations.

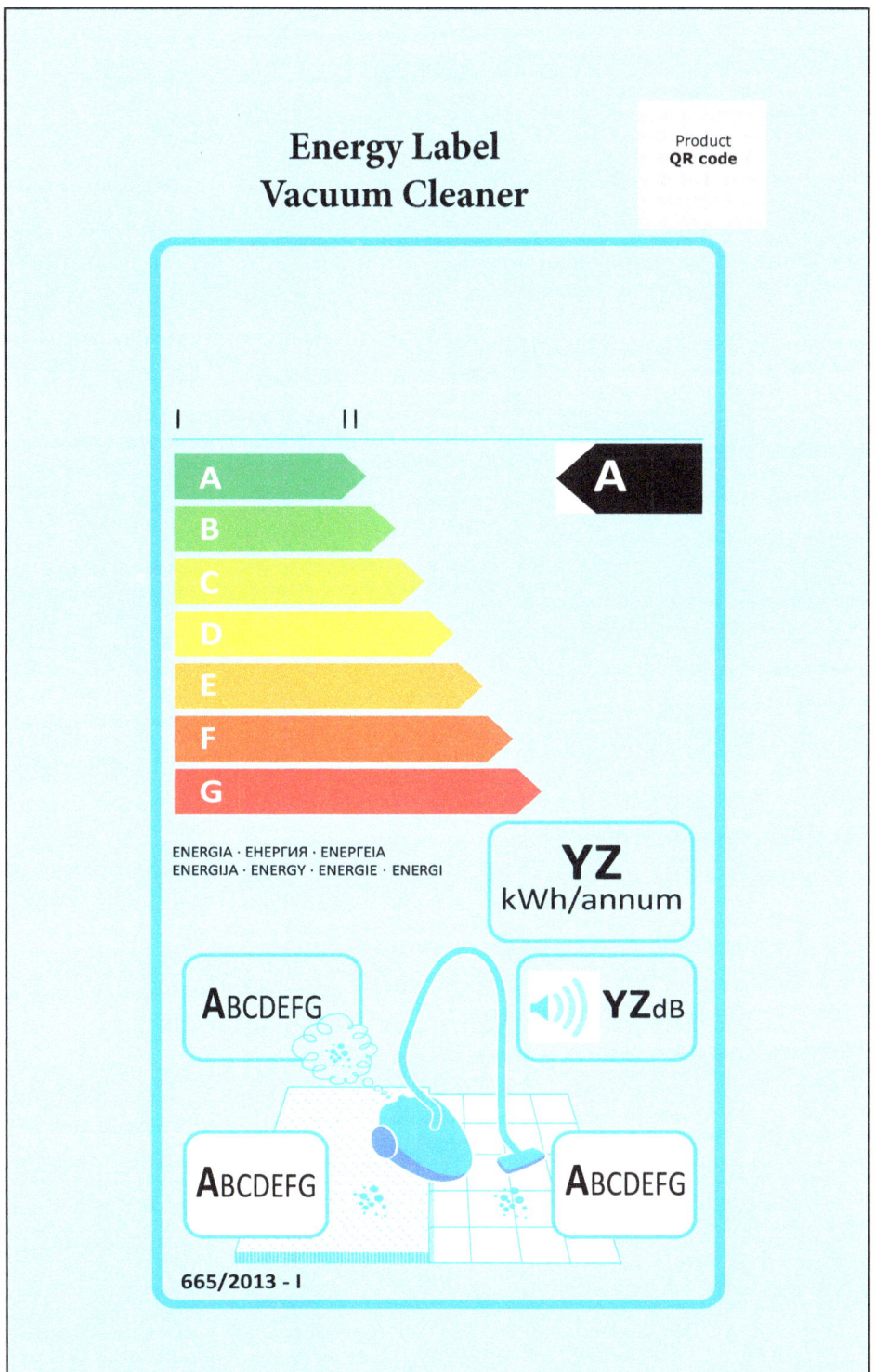

Tyres

Energy label
Tyre energy labels provide a clear and common classification of tyres performance for rolling resistance, braking on wet surfaces and external noise. The labels help consumers make informed decisions when they are buying tyres as they can easily set their priority choice based on the 3 parameters. At the same time, the labels drive manufacturers to innovate to make their tyres appear in the top classes by being more fuel efficient, safer and quieter. The energy efficiency class ranges from A (most efficient) to G (least efficient). A top class tyre has less rolling resistance and therefore requires less energy to move the vehicle. This translates into lower energy costs (fossil fuels or electricity). The wet grip class ranges as well from A (shorter braking distance on wet asphalt) to G (longest). The external noise class ranges from A (less noise outside the vehicle) to B (more noise, with noise levels in the C class not allowed anymore). This noise is different from the "cavity noise", which is the noise transmitted from the rims to the interior of the car.

Energy savings
By choosing a tyre that is in the top class for energy efficiency, and thereby performs the best, energy consumption by 2020 will be up to 45 TWh per year lower than would have been the case without the rules. That is equal to saving roughly 15 million tonnes of CO_2 emissions per year by 2020.

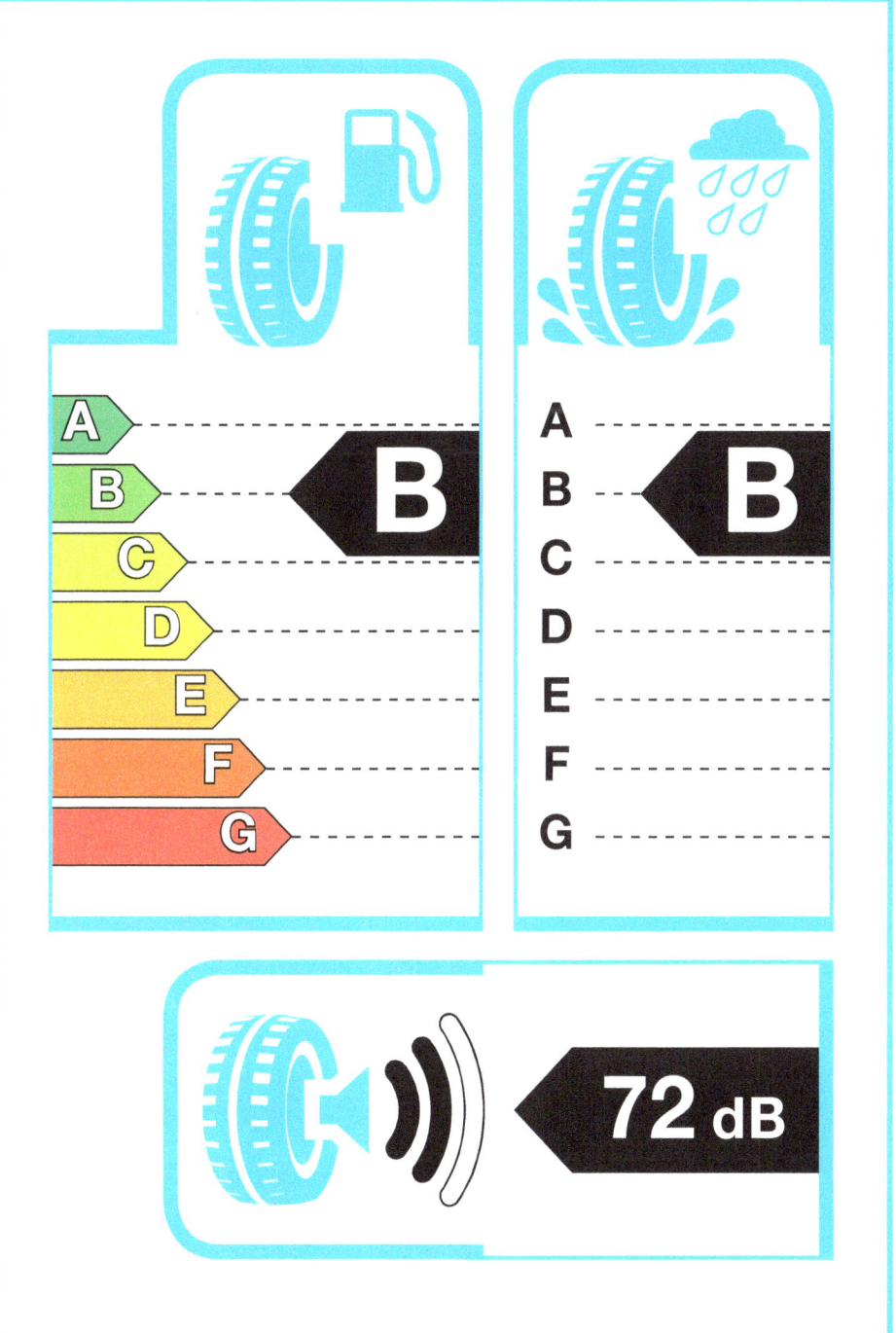

Electronic displays

Energy label

Electronic displays, like televisions, computer monitors or signage displays, are labelled on an energy efficiency scale that ranges from A (most efficient) to G (least efficient). The new scaling system is improved and better takes into account the screen area. The new labels will also show the efficiency of a product when it shows content in HDR, as it can consume twice as much energy as other settings. In addition, the label will also show information on the diagonal size of the display and its resolution, so that consumers can better compare similar displays.

Energy savings
Energy labelling and ecodesign requirements will save up to 39 TWh per year by 2030. Up to 13 million tonnes of CO_2 emissions will also be avoided each year. In addition, 84 thousands of tons per year of plastics will be recycled instead of being incinerated.

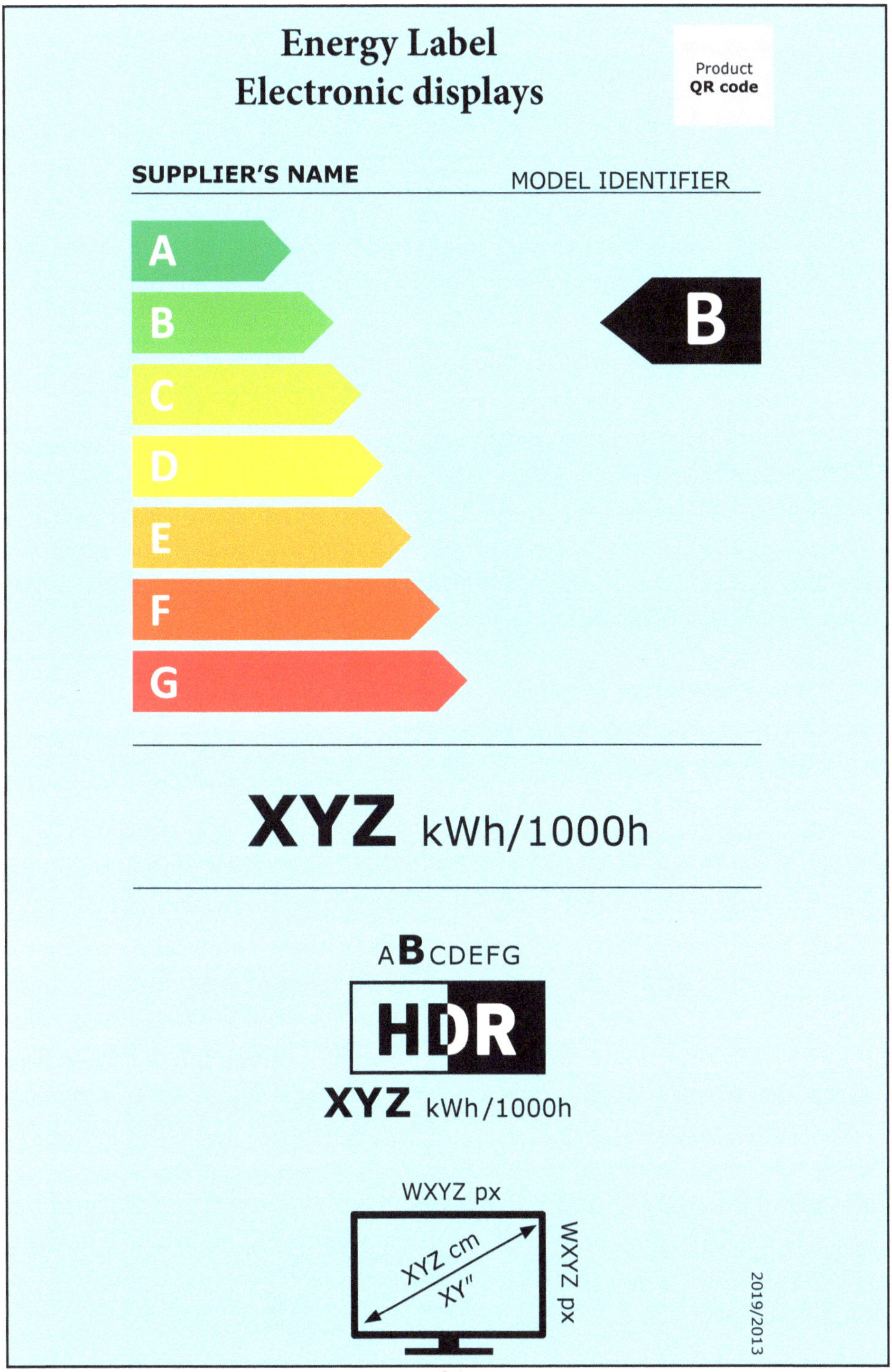
Energy Label
Electronic displays
Product QR code
SUPPLIER'S NAME
MODEL IDENTIFIER
A
B
C
D
E
F
G
B
XYZ kWh/1000h
ABCDEFG
HDR
XYZ kWh/1000h
WXYZ px
XYZ cm
XY"
WXYZ px
2019/2013

Washing machines & Washer-dryers

Energy label

The energy labels for household washing machines and washer-dryers use, a scale from A (most efficient) to G (least efficient). The labels provide information on the product's

- **Energy efficiency class(es)**
- **Energy consumption for 100 cycles**
- **Water consumption for 1 cycle**
- **Duration for 1 cycle**
- **Noise emissions**

Energy savings
Switching to one of the most energy efficient washing machines can save $400 over the lifetime of the product. With more efficient washing machines, Europe will also be able to save up to 1.5 TWh of electricity per year by 2020. This is equivalent to half of Malta's annual final energy consumption and also means a saving of around 100 million cubic meters of water per year.

The new regulations will add to this saving with up to 5 TWh of electricity and 711 million cubic meters of water per year by 2030. This will in turn contribute to reducing greenhouse gas emissions by 0.84 metric tonnes of CO_2 equivalent

Energy Label
Washing Machines

Product QR code

SUPPLIER'S NAME   MODEL IDENTIFIER

- A
- B
- C
- D
- E
- F
- G

B

XYZ kWh / 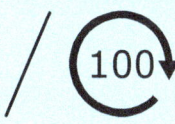 100

 **XY,Z** kg    **X:YZ**    **XY** L

A**B**CDEFG

XYdB
A**B**CD

2019/2014

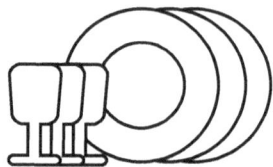

Dishwashers

Energy label

The Energy label for household diswashers uses, a scale from A (most efficient) to G (least efficient). The label provides information on the product's

- Energy efficiency class
- Energy consumption for 100 cycles
- Eco-programme duration
- Water consumption for 1 cycle
- Capacity of the dishwasher
- Noise emissions

Energy savings

By switching to a more energy efficient dishwasher, you can save up to $500 over the average lifetime of the product.

With the implementation of the Energy labelling and ecodesign requirements for efficient dishwashers, the EU will be able to save up to 2.1 TWh of electricity per year by 2030. This will in turn contribute to reducing greenhouse gas emissions by 0.7 metric tonnes of CO_2 equivalent per year.

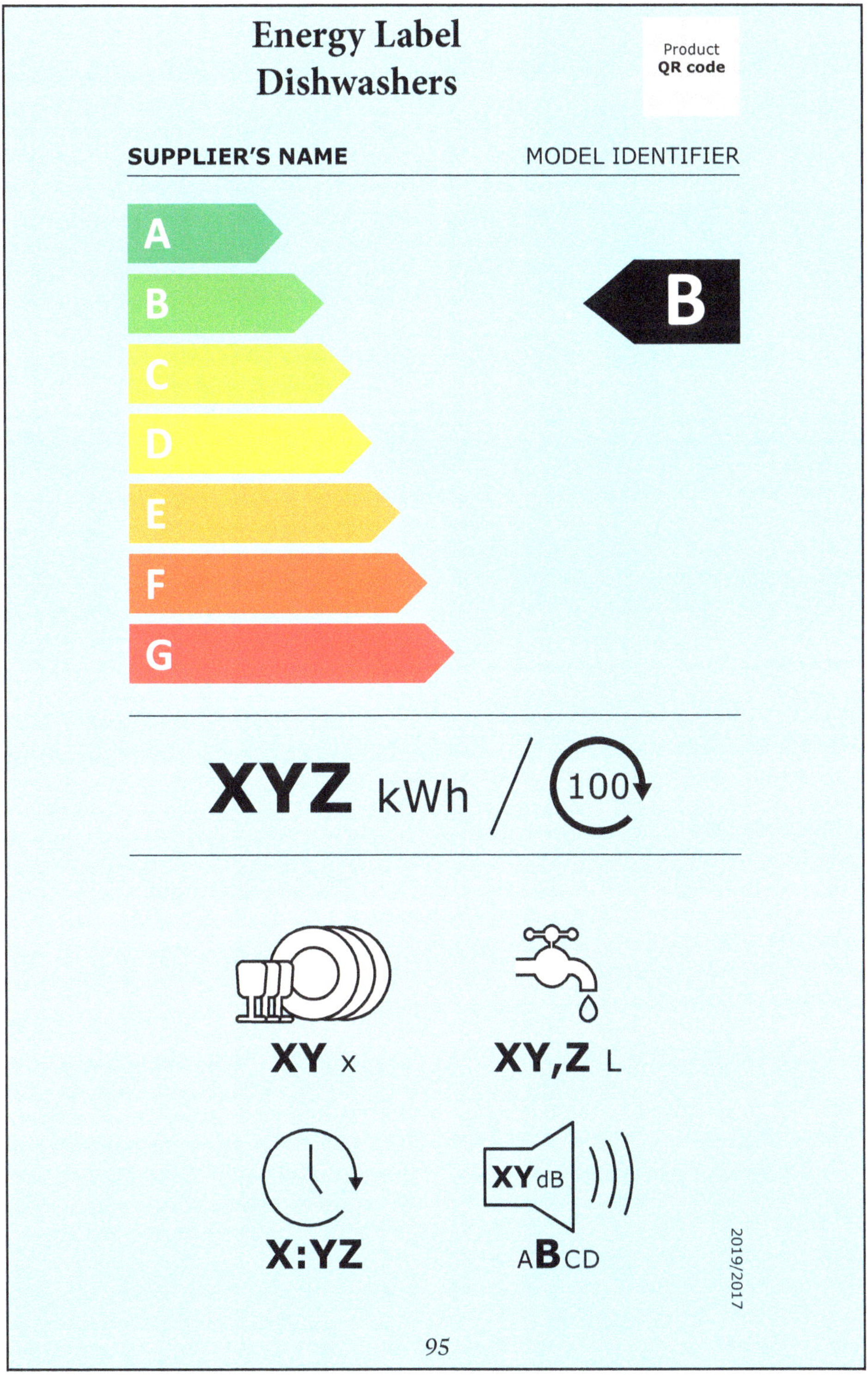

Energy Label
Dishwashers

Product QR code

SUPPLIER'S NAME   MODEL IDENTIFIER

- A
- B
- C
- D
- E
- F
- G

B

XYZ kWh / (100)

XY x   **XY,Z** L

X:YZ   **XY** dB ABCD

2019/2017

Refrigrating Appliances

Energy label

The Energy labels for household fridges and freezers use, a scale from A (most efficient) to G (least efficient). The labels provide information on the product's

- Energy efficiency class
- Energy consumption
- Storage volume(s)
- Whether or not it has a freezer compartment
- Noise emissions

Energy savings

By switching to a more energy efficient dishwasher, you can save up to $500 over the average lifetime of the product.

With the implementation of the new Energy labelling and ecodesign requirements for efficient dishwashers, the EU will be able to save up to 2.1 TWh of electricity per year by 2030. This will in turn contribute to reducing greenhouse gas emissions by 0.7 metric tonnes of CO_2 equivalent per year.

Energy Label
Refrigrating Appliances

SUPPLIER'S NAME MODEL IDENTIFIER

Product QR code

A
B
C
D
E
F
G

B

XYZ kWh/annum

 XYZ L

 XYZ L

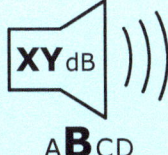

ABCD

2019/2016

Cooking Appliances

Energy label
New cooking appliances come with an energy label showing their energy efficiency class. These range from A+++ to G for range hoods and ovens. For ovens, these ratings are based on their energy efficiency.
For range hoods, in addition to the energy efficiency of the appliance, the efficiency of the extraction, integrated lighting system and grease filtering system is also taken into account.

Standby and off mode
A wide range of electric devices can have standby and off modes.
These are required to switch into a low power mode (such as standby) after a reasonable amount of time and they must not consume more than 0.5 Watts in standby or in off mode.

Energy Savings
By switching to one of the most energy efficient electric ovens, you can save up to $350 over 15 years. With more efficient cooking appliances, Europe will also be able to save around 1% of the annual energy consumed by households by 2030. This means around 2.7 million tonnes of CO_2 avoided annually by 2030 – about the annual emissions of four medium size power plants.

Microwave ovens, outdoor cooking appliances and grills are excluded from these requirements.

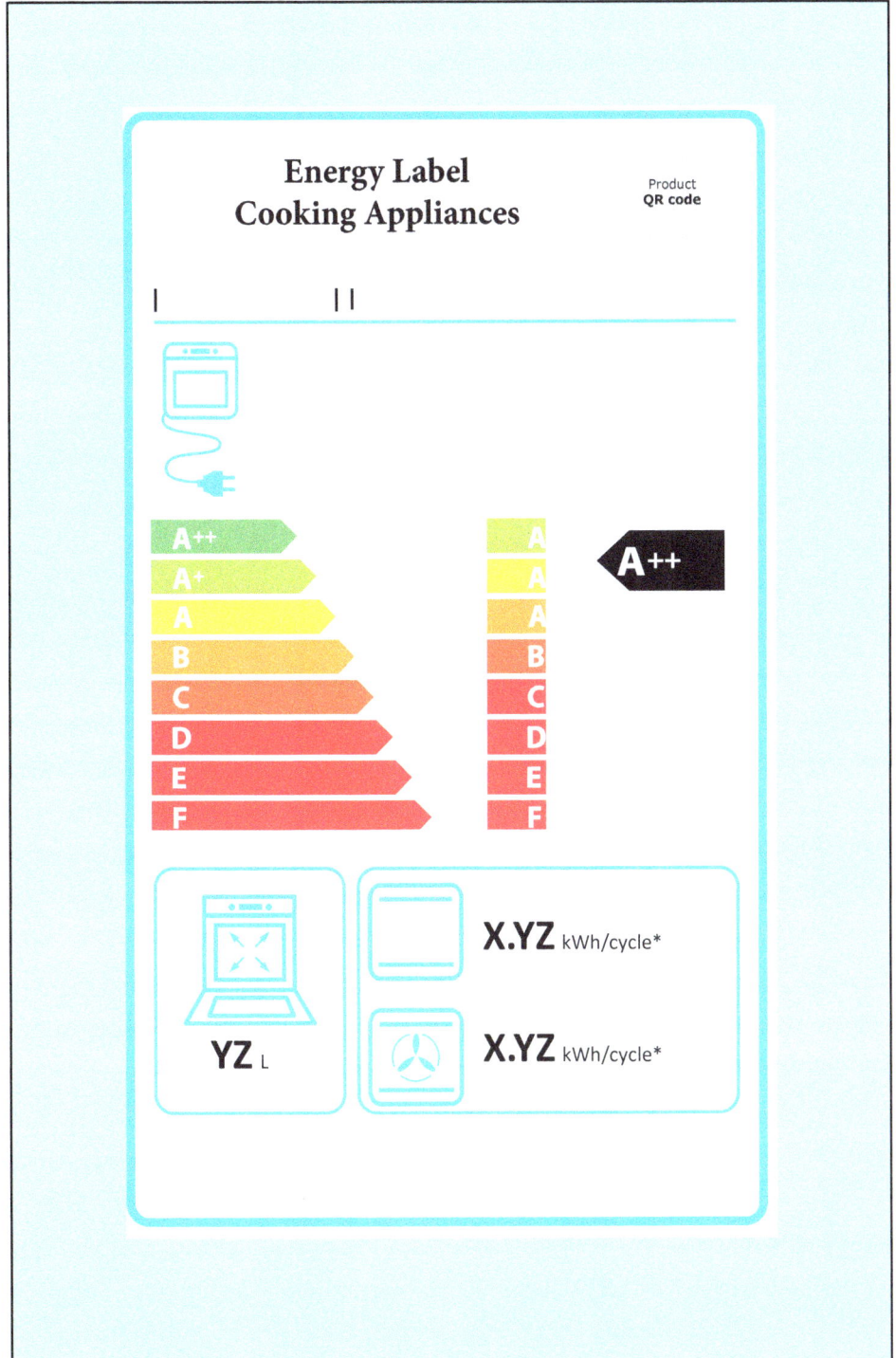

Appendix II

Algae Engine: a Sample for Biofuel promoted by: Top Ten Award International Network

Perhaps, for the first time mentioned, we can use algae for producing water via a very innovative method. At the same time, TTAIN is working on new innovative projects that can effect the future global ecosystem. TTAIN is working hard on the use of algae for preparing enough daily traveling of a vehicle; by using hydrogen produced by algae to store inside the vehicle. The successful completion of this great environmental project makes vehicles not only have

zero emmision but also creates benefits for the environment via producing H2O. So, at the same time, we will be using biofuel with zero emission and producing water that can change the global ecosystem for the better during the years of use. TTAIN hopes that this great innovation will be popular all over the world by **2035** and/or later. The system is very simple because it's just a substitution of the hydrogene tank with an algae stock chamber producing enough hydrogen for these types of vehicles.

APPENDIX III

Environmental Friendly Photos

Environmental friendly photos will be placed in this appendix. These photos can be received in the Top Ten Award International Network inbox from anywhere and everywhere, all over the globe. You can send your appropriate photos to us for them to be considered for publishing in one of the future, related volumes. They will be published with proper credit to the sender. The pictures can also be images of the Ecolabels existing in products within your country.

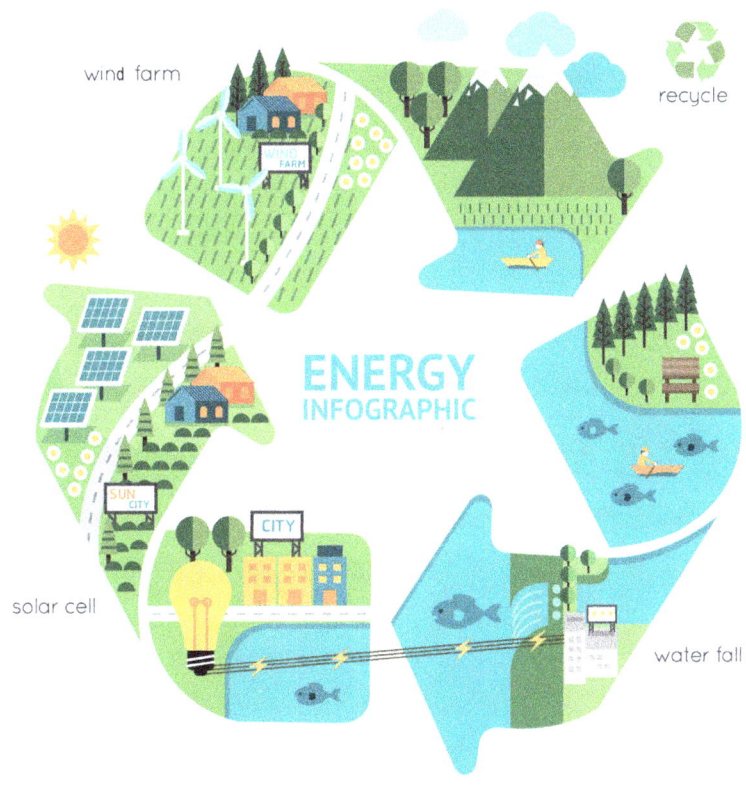

	# Vol.1 Food Industries (Meat, Beverage, Dairy, Bakeries, Tortilla, Grain and Oilseed, Fruit and Vegetable, Seafood, And Sugar and Confectionery)
	# Vol.2 Energy & Electrical Industries (Renewable Energy, Biofuels, Solar Heating & Cooling, Hydroelectric Power, Solar Power, Wind Power, Energy Conservation, Geothermal and Nuclear Power)
	# Vol.3 Fashion & Textile Industries (Fashion Design, The Fashion System, Fashion Retailing, Marketing and Marchandizing, Textile Design and Production, Clothing and Textile Recycling)
	# Vol.4 Health & Beauty Industries (Fragrances, Makeup, Cosmetics, Personal Care, Sunscreen, Toothpaste, Bathing, Nailcare & Shaving, Skin Care, Foot Care, Hair Care and Other Health & Beauty Products)

	**Vol.5** Maintenance & Cleaning Products (All-purpose Cleaners, Abrasive Cleaners, Powders. Liquids, Specialty Cleaners, Kitchen, Bathroom, Glass and Metal Cleaners, Bleaches, Disinfectants and Disinfectant Cleaners)
	**Vol.6** Wood & Stationery Industries (Wooden Products, Cardboard, Papers, Markers, Pens, NoteBooks. Writing Pads and Writing Sets, Pencils, White Papers, Envelopes and Organizers, Staplers and Paper Clips)
	**Vol.7** DIY & Construction Industries (Do it yourself " ("DIY") of Building, Modifying, or Repairing, Renovation, Construction Materials, Cement, Coarse Aggregates. Clay Bricks, Power Cables, Pipes and Fittings, Plywood, Tiles, Natural Flooring)
	**Vol.8** Agricuture & Gardening Industries (Shifting Cultivation, Nomadic Herding, Livestock Ranching, Commercial Plantations, Mixed Farming, Horticulture, Butterfly Gardens, Container Gardening, Demonstration Gardens, Organic Gardening)

	## Vol.9 Professional Products & Services (Teachers, Pilots, Lawyers, Advertising Professionals, Architects, Accountants, Engineers, Consultants, Human Resources Specialist, R&D, Psychologists, Pharmacist, Commercial Banker, Research Analyst)
	## Vol.10 Financial Products & Services (Banking, Professional Advisory, Wealth Management, Mutual Funds, Insurance, Stock Market, Treasury/Debt Instruments, Tax/Audit Consulting, Capital Restructuring, Portfolio Management)
	## Vol.11 Tourism Industries (Airline Industry, Travel Agent, Car Rental, Water Transport, Coach Services, Railway, Spacecraft, Hotels, Shared Accommodation, Camping, Bed & Breakfast, Cruises, Tour Operators)
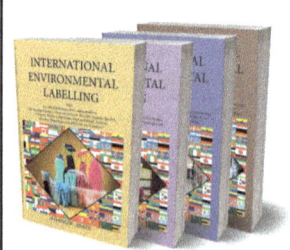	**A Set Box is available that includes Volumes 1-11 as well as a <u>Knowledge Test</u> for all Schools and Libraries, … around the world.**

www.ingramcontent.com/pod-product-compliance
Lightning Source LLC
Chambersburg PA
CBHW040420100526
44589CB00021B/2776